MW01255779

Too Cute!

Baby Deer

by Rebecca Sabelko

BELLWETHER MEDIA
MINNEAPOLIS, MN

Blastoff! Beginners are developed by literacy experts and educators to meet the needs of early readers. These engaging informational texts support young children as they begin reading about their world. Through simple language and high frequency words paired with crisp, colorful photos, Blastoff! Beginners launch young readers into the universe of independent reading.

Sight Words in This Book

a	find	new	white
and	from	the	
are	have	their	
at	help	them	
big	in	they	
eat	look	this	

This edition first published in 2022 by Bellwether Media, Inc.

Library of Congress Cataloging-in-Publication Data

Names: Sabelko, Rebecca, author.
Title: Baby deer / Rebecca Sabelko.
Description: Minneapolis, MN : Bellwether Media, 2022. | Series: Too cute! | Includes bibliographical references and index. | Audience: Ages 4-7 | Audience: Grades K-1
Identifiers: LCCN 2021040715 (print) | LCCN 2021040716 (ebook) | ISBN 9781644875711 (library binding) | ISBN 9781648345821 (ebook)
Subjects: LCSH: Deer--Infancy--Juvenile literature.
Classification: LCC QL737.U55 S2173 2022 (print) | LCC QL737.U55 (ebook) | DDC 599.6513/92--dc23
LC record available at https://lccn.loc.gov/2021040715
LC ebook record available at https://lccn.loc.gov/2021040716

Editor: Amy McDonald Designer: Jeffrey Kollock

Printed in the United States of America, North Mankato, MN.

Table of Contents

A Baby Deer!

Look at the
baby deer.
Hello, fawn!

Fawns are born
in spring.
They stay
near mom.

mom

Most fawns have a **sibling**.

siblings

Fawns **nurse**.
They drink milk
from mom.

nursing

Staying Safe

Fawns hide while mom finds food. They lie in tall grass.

Fawns have
white spots.
The white spots
help them hide.

spots

Growing Up

Fawns find food.
They eat grass
and leaves.

Fawns grow big. **Yearlings** lose their spots.

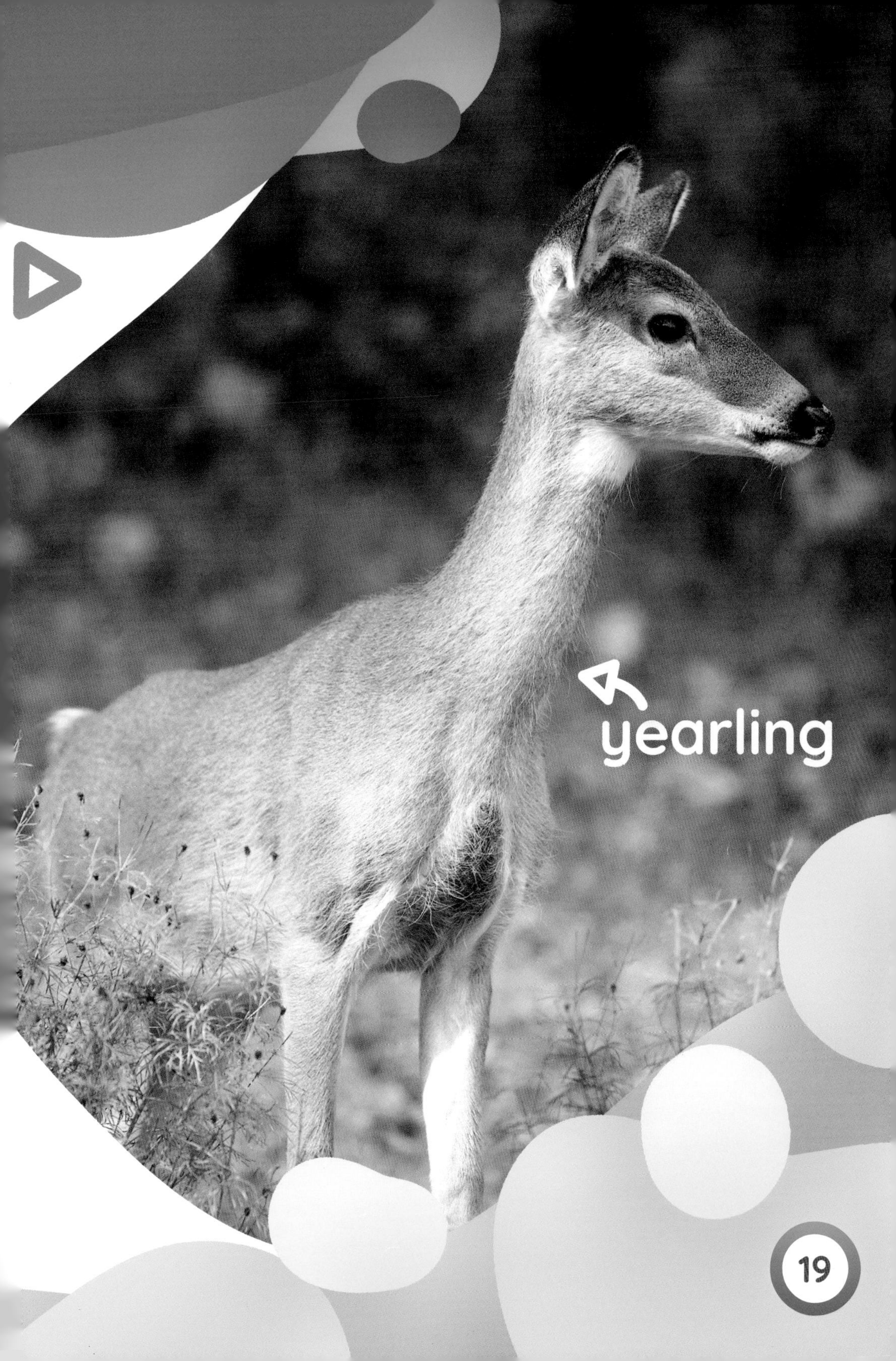

yearling

This yearling finds a new home. Goodbye, mom!

Baby Deer Facts

Deer Life Stages

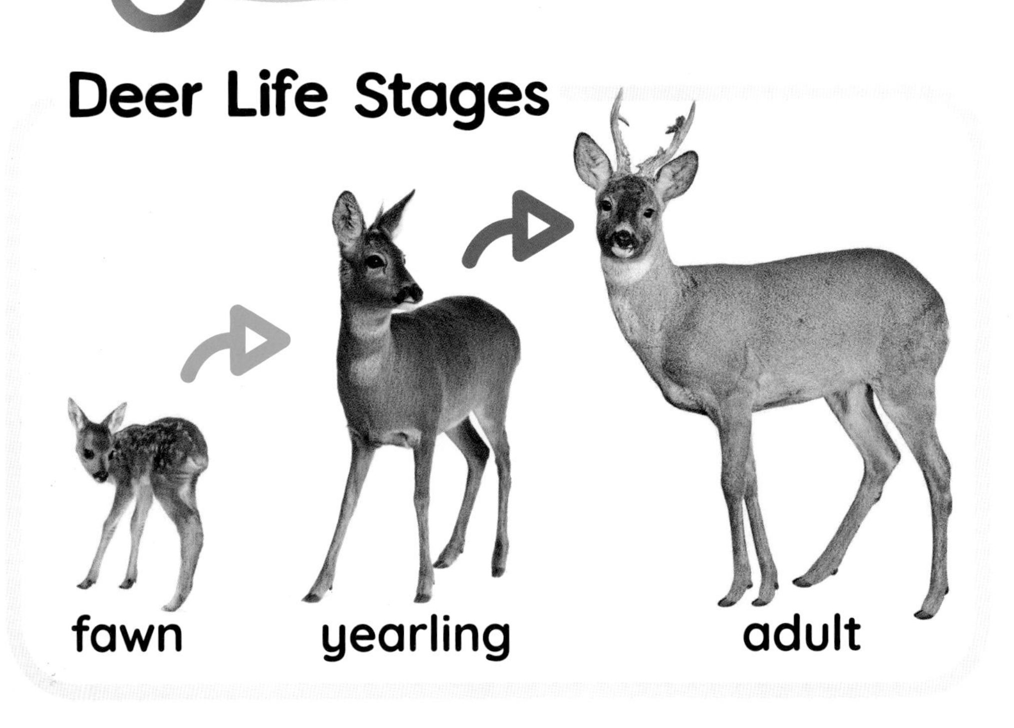

A Day in the Life

nurse

hide in tall grass

eat grass and leaves

Glossary

nurse

to drink mom's milk

sibling

a brother or sister

yearlings

deer older than one year but not fully grown

To Learn More

ON THE WEB

FACTSURFER

Factsurfer.com gives you a safe, fun way to find more information.

1. Go to www.factsurfer.com.
2. Enter "baby deer" into the search box and click 🔍.
3. Select your book cover to see a list of related content.

Index

born, 6
deer, 4
drink, 10
eat, 16
food, 12, 16
grass, 12, 16
grow, 18
hide, 12, 14
home, 20
leaves, 16
milk, 10
mom, 6, 7, 10, 12, 20
nurse, 10, 11
sibling, 8, 9
spots, 14, 15, 18
spring, 6
yearlings, 18, 19, 20

The images in this book are reproduced through the courtesy of: Eric Isselee, front cover, pp. 4, 5, 22 (fawn, yearling); WilleeCole Photography, pp. 3, 16; Geoffrey Kuchera, pp. 6-7; woolerymammoth, p. 8; Victoria Hillman, pp. 8-9; Aurora Open/ SuperStock, pp. 10-11; oumjeab, p. 12; Miroslav Hlavko, pp. 12-13; Melinda Fawver, pp. 14-15; slowmotiongli, pp. 16-17; Guy J. Sagi, pp. 18-19, 23 (sibling); Jim Cumming, pp. 20-21; clarst5, p. 22 (adult); Leene, p. 22 (nurse); Rudmer Zwerver, p. 22 (hide); Arunas Solovjovas, p. 22 (eat); Linda Freshwaters Arndt/ Alamy, p. 23 (nurse); WildMedia, p. 23 (yearling).